happier together

HOW TO FIND YOUR PEOPLE
& MAKE FRIENDS THAT LAST

Dr. Lori Whatley

hatherleigh

Hatherleigh Press, Ltd.
62545 State Highway 10
Hobart, NY 13788, USA
hatherleighpress.com

HAPPIER TOGETHER

Library of Congress Cataloging-in-Publication Data
is available.
ISBN: 978-1-961293-21-2

Printed in the United States

The authorized representative in the EU for product safety and compliance is Catarina Astrom, Blästorpsvägen 14, 276 35 Borrby, Sweden. info@hatherleighpress.com

10 9 8 7 6 5 4 3 2 1

To my darling grandson, Fischer.
May you know the joy friendship brings.

"If you have one true friend you have more than your share."

—Thomas Fuller

CONTENTS

"In the sweetness of friendship
let there be laughter and sharing
of pleasures. For in the dew of
little things the heart finds its
morning and is refreshed."

—Khalil Gibran

THE BASICS OF FRIENDSHIP

"Some people go to priests, others to poetry; I to my friends."

—Virginia Woolf

Once upon a time, there was a lady named Mary. Mary was 102, and people came from near and far to study her, eager to find out how she'd managed to live such a long life. After all, isn't that everyone's goal: to live a long, healthy life? Everyone who spoke with Mary came to realize there was one element of her life which they felt was responsible for her centenarian status: friendship. *Mary not only had a collection of close, personal friends, but she was also a friend to others. This was the "secret sauce" in Mary's long life. Mary had friends in high places, low places...all places!*

And together, they were on the journey of life, sharing and caring and shining light on each other's path to find the way. Mary truly knew the value of community, personal connection and spending one's life with others.

Loneliness is the opposite of connection, and it is as hazardous to our well-being as smoking 15 packs of cigarettes a day. Being disconnected and isolated places a physical and mental toll on our body that is equivalent to drinking six alcoholic drinks a day. As humans, we are socially interdependent, meaning we *need* each other to thrive. Who knew?

The answer is, we *all* knew—and if we only paid more attention to the way we feel in moments of loneliness, we would naturally move toward our intrinsic need for community. Yet isolation continues to be a universal problem, with much of the population completely unaware of why we are suffering from epidemics of anxiety and depression at a higher rate than ever before.

Social isolation is now one of the biggest risk factors to our health, with research finding that not only can we die lonely, we can also die *of* loneliness. In other words, a part of any healthy lifestyle is connection and it is just as important as regular exercise or avoiding smoking. The top predictor of living a long life is having close relationships and interacting with them daily.

There is value in using the friend connection in every stage of life to create a strong foundation and live our best lives. At each stage of life, having friends and being a friend brings us so much comfort that researchers are even claiming that a life full of meaningful connections is the very best way to do life.

Despite the results of this research, look at what's happened to us. Nowadays, we are disconnected and lonely. The ways in which we interact with one another are impersonal, ineffective. Nothing we do satisfies our craving for human connection. We are not living true to the way we were created—to be connected and engaged with one another.

Meet Val. All of his happiest stories include his friendships. The difficult times he has endured in life have always been manageable thanks to his having close friends by his side. He believes community empowers us. Val treasures his friendships and knows the importance of being a dear companion. As he has aged, he is always surrounded by friends; living a solitary life has never crossed his mind. One of the very best gifts of life is to be loved by another and to share life's highs and lows with a friend. So, his best advice is to find connection and friendship to enrich life and be happy. Sharing is caring.

When we try to uncover the secret to a longer life, a more satisfying life, again and again we find that the secret is friendship. For centuries, we have touted the foods we consume or proper amounts of sleep and exercise as characteristics of a healthy life, but now we understand friendship to be one of the most important benchmarks for longevity. Companionate relationships foster longevity. When we build friendships from mutuality, these relationships become more than a source of happiness—they also provide the support we need for overcoming life's challenges. In short, it is not a lack of love but a lack of friendship that makes our lives unhappy and often seem unmanageable.

GET CONNECTED

"A friend is someone who makes it easier to face the challenges of life and helps you to make more of the good times."

—Barbara Bush

Healthy connections with other humans (along with a need for community) is part of our essential construction. Humans are hard-wired to interact with one another. That's why, when I see a client who is depressed or anxious, one of the areas I look into first is their social life. If we aren't connected, we are more likely to be depressed, lonely and anxious. The good news is that it works the other way around, too. Establishing new connections with other humans is a great way to avoid depression, loneliness and anxiety. Make connecting with others an intentional part of your life, and you will start to notice an uplifted mood and a more peaceful outlook.

WHY DO I NEED FRIENDS?

"I would rather walk with a friend in the dark than alone in the light."

—Helen Keller

In this book, my mission is to convey to you the importance of friendship, the requirements for being a valued friend and how to develop your own friendships. These are challenging propositions, but I believe that once we understand the benefits on offer, we will be more willing to endure these difficulties to better enjoy the pleasures of good friendship connections.

The first step is understanding that, in order for real connections to take place, in-person interactions are essential. Lasting relationships cannot be formed solely through text messages or from behind a screen. These types of relationships are well worth the effort, however: they are precious and add value and meaning to our life's journey, just as we add the same to our friend's life.

If our life is to be a pathway lined with mutual loving and caring, friends are an essential need. Friends are the people who walk in the door when difficult times come around while everyone else is walking out. They're the ones who sit with us without saying a word, yet their mere presence is all the comfort we need. They celebrate with us when we land that new job, buy the first house, earn a big promotion. Whether they have been with us the whole way or they're new to our lives, old friends and new are equally precious, adding meaning and value to our life's journey.

So, we understand that friendship is the antidote to feeling disconnected and lonely. It is a simple formula for a long and lovely life of contentment: cultivate friendships with people who love us and love them in return. This is the basic formula for meaningful friendship and authentic success, life's greatest treasure.

But how *do* we make friends?

Happier Together will help you understand the importance of connections and the best ways to create them—the whys and hows for a life of contentment. We will also examine how and why we've become so disconnected and investigate avenues for reestablishing connections with others. By learning best practices for finding your people and walking with them on this journey called life, you can avoid the pitfalls of broken or unbalanced personal relationships.

To address our own disconnect, we first need to notice the habits that we have which disconnect us. We have become confused; we think friends met online are the same as in-person friendships. The truth is, they are not. Online we don't receive the eye contact and human touch required to activate the pleasure chemistry into our brains. Likes and follows are not the same as hugs and physical touch and cannot play the same role as an antidote for depression and anxiety.

So, let's start by identifying what we are looking for in a friend, the types of friendships available to us, and what we can do to start developing these relationships and obtaining happiness through connection.

PART I

HOW TO FIND YOUR PEOPLE

"There are no strangers here.
Only friends you haven't met yet."

—Thomas Aquinas

THE CONNECTIONS WE ALL WANT

It cannot be understated how important connections are for our well-being. We were created to connect to and be in relationships with other humans. Having intimate friendships gives us a 50 percent higher chance of overcoming struggles, while people who are habitually lonely with no social connections have an increased risk of premature death. Research has even discovered that we are unconsciously drawn to people who challenge us in the exact ways that we need to grow.

This revelation offers us a new perspective for understanding the importance of meaningful social connections. If we are willing to look at connection as an opportunity for growth and to make the changes needed to reform unhealthy patterns in ourselves, then friendships don't just fulfill a basic human need—they offer us a chance to learn and improve.

INTUITIVE CONNECTIONS

In regards to making the right connections, we often work from a place of instinct. When we have met the right type of person the signs are all there and we get the internal green light. Sometimes the evidence is obvious: the people who celebrate your wins and support your goals, who listen to what you have to say, who make you feel safe. People you don't have to weigh your words with and who respect your boundaries—these are *your* people. Your safe connections. Trust your heart to choose your friends.

Birds of a Feather

It's interesting to stop and notice the friend group you have created. Experts say that friend groups are typically composed of like-minded individuals. Birds of a feather flock together, in other words. This works in the other direction, too: if you intentionally put yourself in situations with people you admire and want to be like, you are more likely to develop new friendships—and those friendships are more likely to be authentic, supportive and long-lasting.

WHAT TYPE OF FRIEND ARE WE LOOKING FOR?

"There is no exercise better for the heart than reaching down and lifting others up."

—John Holmes

The first step in developing new, long-lasting connections lies in identifying which kinds of friends we need. For example, consider the type of friend who shows up when everyone else is walking out the door. Experience teaches us that when hard times come around, many friends make the choice to scatter. The good news is we only need three or so close friends who will sit with us in our mess. These are our people, our ride-or-die friends; all the rest are simply surface friendships. Surface friendships also serve an essential purpose, of course. Each type of friendship plays an important role in our life and contributes to it in different and significant ways. We need all kinds of friends, and fostering friendships with all types of people can only be to your benefit.

Variety is the spice of life, and nowhere is that more relevant than in the friends we spend our time with. While it may be true that "birds of a feather flock together," there's room in our lives for all types of friendships, each meaningful in its own way.

Each friendship is special, and each friend plays a different role in our lives. Consider:

- **Do you have a friend who listens to you?** Someone who is always there to lend an ear and offer support when you need to talk about your feelings and problems, wins and losses?
- **Do you have a friend who advises you?** Someone who always has wise counsel and guidance to offer, affording you valuable perspectives and helping to make difficult decisions?
- **What about a fun-loving friend?** A friend knows how to have a good time and brings laughter and joy into your life, who can turn a rainy day into a grand party?
- **Do you have a friend who challenges you?** We all need a friend who pushes us out of our comfort zone, encouraging personal growth and helping us reach our full potential.

- **What about a loyal confidant?** The friend who is trustworthy and keeps your secrets safe, providing us with a sense of security and trust in our relationship.
- **Do you have a friend who cheers you on?** Someone who celebrates your successes, cheers for you during tough times and boosts your confidence?
- **What about an honest critic?** The friend who provides constructive criticism and feedback. These friends don't sugarcoat things, and this helps us improve and grow.

Having a diverse group of friends who each fulfill unique roles in our lives not only enriches us; it provides us with many types of support in the different life arenas.

The Great Unfolding

Author Robert Louis Stevenson once said, "A friend is a gift we give ourselves." Many do consider a friend a gift that makes life more manageable and enjoyable. Moving through the stages of friendship is like opening a wrapped gift, peeling back the paper to discover what's inside.

The period of time during which a new acquaintance begins to bloom into a more special connection—a real friendship—is a beautiful, memorable and unique experience. The gradual realization and deepening of understanding, connection and trust between two friends over time is an incredible journey of discovery as you each uncover the other's true self. You learn to respect each other's strengths and overlook each other's faults while creating and nurturing a bond that will withstand life's ups and downs. By supporting one another's growth and finding comfort and joy in each other's presence, we exchange this special gift unique to human beings which makes the journey of our lives that much sweeter.

TYPES OF FRIENDSHIPS

Not all friendships are created equal, which leads to our acting differently with different people. This isn't deception or concealment; we simply show different sides of ourselves depending on the situation and the people we're with. Work friends will bring out one behavior style; school friends bring out another. We can be silly with some folks but serious with others. We have a role in each connection and that's perfectly normal. In other words, we will have different types of friendships with different types of friends.

SOCIAL FRIENDS

Social friends are best described as those we share social activities, interests and experiences with, but with whom we may also have a more defined connection as compared to close friends or family members. These are the people we enjoy spending time with in social settings such as parties, group outings and other events. When we are with our social friends, we feel inspired. They show up when they are needed most, motivate us to take action, help us feel secure, give us space to be unique and bring out the best version of us. When we find friends like this, we know we have found true companions for our life journey.

In social friendships, we can simply pick up where we left off even when we haven't seen each other in a while. This is because social friends are the people who choose us and whom we choose. Balance and reciprocity are at the heart of social friendships which makes them easy to manage. They are invested in connection through the things that matter to us and are committed to working together to make our lives healthier and happier.

Social friendships provide companionship, support and a sense of belonging. They offer opportunities for companionship and fun through sharing common interests. Additionally, these sorts of friendships can provide networking opportunities while offering different perspectives which help us learn and grow as individuals.

The choice to have social friends can increase our happiness and improve our health. It's important not to get so busy with life that we don't foster this element of friendship to enrich our lives. For example, I play Mahjong on Tuesdays and Bunco on Wednesdays; the people I play with are social friends. I also play pickleball on Fridays with social friends. We enjoy a super club on Friday evenings—social friends again. We don't have to discuss anything too heavy, though we can if need be. These are people we connect with because we enjoy socializing with them. There are no other connections; we are not related, we don't work together, we are not super close. These are more surface friendships, valuable in their own right and essential for a well-rounded social existence.

There are certain practices you can integrate into your social friendships to increase your chances of forming more meaningful and lasting friendships. For example, offering kindness and reliability is a real builder of friends. We all enjoy those elements of friendship which feel supportive. Understanding that connections take time to build and being persistent in our practices is encouraged. Consistent efforts like following up after meeting someone new can take a social friendship to the next level.

Three is the Lucky Number

A wise mentor once told me we should have three close friends, and the rest should be surface. Maintaining more than three close friendships can be exhausting mentally as well as physically. I've found that structure beneficial, to the point where I teach it in workshops about friendships and healthy lifestyle balance.

INTIMATE FRIENDSHIPS

Intimate friendships go beyond the limitations of social friendships—requiring more investment and providing greater returns. These are deeper and closer relationships, involving emotional connection, mutual trust and vulnerability. They go beyond surface-level interactions and involve sharing personal thoughts, feelings and experiences with one another.

Often, our most intimate friends are also our oldest friends, those who we've known the longest and shared the most with. Old friends are like fine wine—they get better with time. Having weathered storms together, you've built a bond that withstands the tests of time and hardships.

There is high value in this sort of friendship. It's irreplaceable because of the deep shared history, understanding and connection: old friends have seen each other at their best and worst and still choose to be at each other's side. These friends know all of our secrets and still love us, and our friendships with them carry years of memories and shared experiences that enrich our lives in so many ways.

It is within these deep friendships that we learn things like how to deal with others, who we are currently, and who we'd ultimately like to be.

Our Furry Friends

"Until one has loved an animal, a part of one's soul remains unawakened."

—Anatole France

Healthy friendships aren't formed solely by humans connecting with other humans. Our pets can be one of our greatest sources of joy and companionship. Whether we are playing fetch with our dog, cuddling with our cat or simply watching the fish in our aquarium, the joy this brings to us is amazing. This is because the bond we share with our pets is truly special. We learn from them about unconditional love, loyalty, and the importance of living in the moment. In a hectic world, spending time with our pets can be a gentle reminder to slow down, appreciate the simple things and cherish the connections we have.

Of course, connections with humans contrast greatly from our relationship with pets and are clearly not substitutes for human connections. The nature of the interactions and the depth of connection are very different indeed. Humans can provide emotional support through communication, understanding and reciprocity in ways our pets cannot. Humans have evolved to form intricate social connections that fulfill a wide range of needs—emotional, psychological, intellectual and social—which pets, despite their wonderful qualities, do not fully replicate.

WHERE TO FIND OUR PEOPLE?

We have identified the different types of friends who help create diverse connection types. Hopefully, you can look at your friendships and identify at least a few of these archetypes in your group. However, if you don't already have these friends, how can you acquire these types of connections in your life?

When forming friendships, mutual interests are often the glue that holds a nascent connection together while it grows and develops into a stronger bond. That being said, opposites really *do* attract one another—which works out well, as opposites also complement one another. Don't shy away from cultivating friendships with people you have nothing in common with. You will be surprised how fun it is to learn new things from friends who are different than you.

In looking for new friendships, however, rejection is always a possibility. While never enjoyable, it's important to realize that rejection is often about the other person, and is not an indictment of you as a person (or as a potential friend). It is essential for you to move on and find better fits and other friendships. Not all people will be our friends, and that's *fine.* We are just looking for a few solid connections, and if we embrace the search for friends as an adventure all its own, our perspective shifts from fear to fun.

With all that said, let's consider some of the different places that we can meet new friends.

ONLINE FRIENDSHIPS

Nowadays, one of the easiest places to find friends is online. Social media, online games and internet forums can provide us with hundreds of potential connections a day, but we are seeing more and more that not everything we see online is as it appears to be. Online friends have the unique ability to deceive, appearing one way when they are actually another. It is also fairly easy to disappear from someone's online life instantaneously and completely—delete, block and disappear.

The modern trend of "ghosting" someone is just one frustrating aspect of friendships that exist solely online.

Online friends are also hard to really get to know and thus form friendships that are not structurally solid. Some are very temporary, exacerbated by the types of communication common to online-only connections. For example, text messaging is a major component for most digital friendships. Although instant messaging can be a helpful way to maintain friendships, we must be mindful that having serious conversations through texting can be disastrous for relationships due to a high risk of misunderstanding someone's true meaning. Humans rely on so much more than words alone to convey meaning to one another—we're reliant on facial expressions, body language and tone of voice to get our points across. Stripped of these communication tools, we're left trying to put together a jigsaw puzzle in a pitch-black room.

There's also the issue that online friendships serve as a sort of instant gratification for human connection. Say we've had a bad day; we go online to unload and quite often we'll find one or two kind souls willing to respond and offer comfort. This feels wonderful in the moment, as it satisfies an immediate desire, but does nothing to help us grow or develop. If these are the only sorts of friendships you have, it may be time to actively seek out some in-person friendships—ones where you can sit with each other, look into one another's eyes, offer healthy physical contact like big hugs. This is paramount in our human needs structure, as Maslow's Hierarchy of Needs so brilliantly points out.

Abraham Maslow was a psychologist in 1943. He wrote a now infamous paper titled "A Theory of Human Motivation." In it he discussed the different types of human needs paramount for mental health. The needs he discussed in this paper are the basic requirements for human survival. The need he placed as number three in the hierarchy is the need for love and belonging. Humans have a desire to be a part of a community or social group and to enjoy affection in these relationships in ways that can only happen in person.

Having pointed out the downfalls of online friendships, we can also look at the ways they work. First, online friendships are a great place to start. They are a beginning in terms of connection. We can find new friends here and then foster the friendships in person for deeper growth. We know that there are many emotional and physical needs that can only be met through in-person interactions with other humans.

We can be mindful to reach out regularly online to maintain and enhance friendships. We can use messaging apps, social media and video calls to stay connected. Being responsive and timely in our replies shows that we value the friendship and are making an effort to keep the relationship growing. In these ways, we can nurture healthy online friendships that bring positivity, support and joy into our lives.

Here are a few more tips on how to ensure that online interactions serve as tools rather than substitutes for genuine relationships:

- **Use platforms for purposeful communication and to facilitate deeper conversations.** Meaningful conversations and support are key components of friendship that can be maintained online.
- **Balance real-life and online interactions.** While online interactions are convenient, make sure to balance them with offline interactions to build stronger, more meaningful friendships.
- **Focus on the quality of interactions rather than the quantity.** Maintain connections with existing friends rather than replacing face to face interactions.
- **Engage in shared activities online that strengthens bonds** such as playing games together, collaborating on projects or discussing mutual interests.

- **Use online platforms to organize face to face meetings whenever possible.** Physical interactions help solidify friendships and create lasting memories.

By approaching online interactions with intentionality and using them to complement rather than replace face to face friendships you can ensure that they serve as effective tools to maintain and nurturing meaningful relationships.

DISCONNECT TO CONNECT

It's interesting that many of us tend to interact with our online friends more than we do with our in-person friendships. The Pew Research Center has found that online friendships lack critical components—like physical presence and face to face interactions—that are integral in building deep connections. So, if you want a deep connection, you'll need to first disconnect from your digital devices to find it.

We can't substitute online friendships for in-person connection. On top of the shortcomings of online-only connections, there's a tendency to think that online interactions fulfill our basic need for human contact. People will self-isolate, particularly in this new work-from-home era, believing they are safe from the major repercussions of solitary in-home living just because they are talking to people online.

Let's face it: a "like" is not the same as a genuine smile or a walk on a sunny day outside with a friend. Unfortunately, our brains react to likes by providing us with a hit of dopamine, a pleasure chemical like we would receive from eating a piece of chocolate. These spikes of satisfaction keep us coming back to social media for our hit of happiness, but these "likes" are too shallow to create solid friendships. The lack of commitment inherent to online interactions encourages friendships that dissolve easily. It's just not the same as an in-person friendship interaction...even if our brains trick us into believing otherwise.

Which isn't to say online interactions have no value—they provide a large quantity of opportunities for meeting and maintaining friendships. The trick is not to treat them as a substitute for real, in-person experiences together. For example, I recently met a new friend online who graduated from my alma mater. She is a performer and, when she came to my city, I disconnected online to attend her performance in person. This is a great example of leveraging online experiences to add to, rather than take away from, our lives.

RED FLAGS AND PRACTICAL WAYS TO AVOID DANGERS ONLINE

It is not uncommon that we will encounter online unpleasantries. We must be aware of the possibilities and know what to look for and to avoid when interacting online. One such common online danger is unsolicited requests for personal information. Be cautious of individuals on websites asking for personal information such as your address, social security number or financial details without legitimate reason for this.

We should also be wary of requests for money or financial assistance from individuals we meet online, especially if the request is accompanied by an urgent plea or sob story.

If someone seems overly familiar or pushy when reaching out to you with unsolicited messages or friend requests, be concerned. If you feel pressured to move a conversation to a different platform, for example from a dating app to in-person texting or email, be cautious.

These dangers can be avoided using practical tools such as privacy settings. Think before you post personal stories, photos or location details online. Trust your instincts if something feels off or too good to be true and proceed with caution. By staying vigilant, informed and proactive, you can navigate online interactions safely and minimize the risks associated with potential dangers.

STRANGER DANGER

Another concern with online friendships is the safety factor. We must approach online friendships with caution, ensuring safe and healthy interactions through set boundaries.

We've all been taught how sharing personal information online carries risks of privacy breaches and exposure to online predators. Others report experiencing cyber-bullying and online harassment, which negatively affects their self-esteem and peace of mind. And even if you avoid these obvious dangers, there's a new keyboard warrior born every minute, and online conversations make it easier than ever to be aggressive and verbally hostile in ways you'd almost never see in-person.

COUCH POTATO

When we spend too much time online, we risk social isolation and neglect of our real-life social connections and activities.

Isolation is bad for our mental and physical health, with a wealth of research having been done into the negatives of a digital device-based life vs. an in-person, play-based life. If we are outside in nature, enjoying the sunshine and listening to birds sing, it is much better for our mental wellness than being inside on digital devices for hours each day. Yet shockingly, people spend an average of eight hours online each day, resulting in a harmfully sedentary lifestyle which we know contributes to mental and physical health concerns such as cardiovascular disease and diabetes. Anxiety and depression have also skyrocketed in adolescent age groups which are consistently online.

Online interactions are here to stay, and the issues attendant to them will only worsen as metaverses and online spaces become more prevalent and advanced. For these reasons, it is essential to discover healthier ways we can harness the benefits of technology while still prioritizing meaningful human connection. This means being intentional about striking a healthy balance between online and in-person interactions. We must commit to navigating the digital landscape mindfully and working to foster authentic relationships in an increasingly virtual world.

There are ways that we might enjoy technology while also building our in-person relationships which is crucial for maintaining a healthy lifestyle.

- **Set boundaries and establish specific times when we will be offline.** Focus on in-person interactions, for example, during meals or on certain evenings.
- **Learn to use technology to enhance rather than replace.** For instance, we use tech to set up in-person meetings or sharing experiences rather than substituting them. If we limit our screen time, we can be certain it doesn't interfere with face-to-face interactions.
- **Practice active listening when spending time with others.** Start by giving

them your full attention, without distractions from phones and devices. There are many tech-free activities we might engage in that won't involve screens such as outdoor adventures, sports or board games.

- **Prioritize quality time by investing in meaningful conversations and shared experiences.** We do this by avoiding social media or comparing our relationships to idealized versions seen online, and instead focus on authentic interactions to build deeper connections.

Leading by example looks like modeling healthy tech habits for those around us, especially children and younger family members. Designating tech free zones in your home like dining areas or bedrooms also encourages face to face interactions. We must regularly reflect on how tech use impacts relationships and then adjust accordingly to maintain a healthy balance.

By integrating these practices, we can harness the benefits of technology while nurturing and prioritizing our in-person relationships.

"Pal 9000"

It should go without saying that artificial intelligence (AI) cannot replace human connection. You can't connect with another person if there's no person to connect with, after all! Apps that allow you to carry on full conversations with AI characters are certainly an interesting novelty, but the power of human connection is so much more powerful and complex than a computer could ever simulate. We *can* utilize AI to make our lives easier, freeing up more time to be together with the important people in our lives. Never forget that AI is a computer, not a companion!

WORK FRIENDSHIPS

Making friends at work can greatly enhance your overall work experience, productivity, and job satisfaction. However, it can be challenging to transition these workplace connections into more meaningful, wider ranging friendships. Let's look at some ideas for making deeper connections with our professional colleagues.

First, we must be willing to begin conversations with our coworkers. This could take the form of you showing interest in how their weekend went or discussing how a particular project is moving along at work. This also provides us an opportunity to practice our active listening skills. Listen as they share about their lives, paying attention to what they're saying rather than only listening to comment afterwards. You will be amazed at how receptive people are to talking about their lives and sharing their experiences. Let's face it, we all enjoy talking about our hobbies, family and recent activities we've enjoyed. Sometimes, even just smiling at the other person can activate oxytocin, the bonding chemical in our brains.

From acts as simple as this, friendships are born and grow through times like this spent together, sharing and listening. We all appreciate a friend who is genuinely interested in us.

MAGNETIC PERSONALITIES

Are you approachable? Being easy to approach and talk to will greatly increase your chances of forming secure connections in the workplace. By smiling often, making good eye contact and giving thoughtful responses to questions asked by coworkers, we are giving others the greenlight to engage, which improves our chances for closer relationships.

We should be able to easily find common ground with those we work with. It may be an interest in the same sports team, a shared hobby, or similar backgrounds. Once we are engaged and connected, we can enjoy these commonalties as they encourage the unfolding of friendship. But the workplace also offers unique opportunities to help one another on different tasks. Whether that means offering advice or lending an ear, this cooperation facilitates connection to strengthen friendships in the workplace.

Ask a trusted friend for their opinion of you. In what ways are you approachable? What are some areas you can work on?

How can you be more approachable? Being more approachable involves a combination of body language, communication style and attitude. Here are some tips to improve your approachability.

- **The easiest one is to smile.** A genuine smile is universally inviting and shows you are open to interactions. Smiling can also benefit you by giving you a subtle dopamine boost which improves your mood.
- **Avoid crossing your arms and legs.** This can signal defensiveness; instead, keep your posture open and relaxed.

- **Make friendly eye contact with others.** You should also avoid staring, which can be intimidating.
- **Pay attention to what others are saying through active listening.** Don't interrupt and instead show interest in their thoughts and feelings.
- **Use positive gestures like occasional nodding to emphasize points.** Mirroring the other person's body language subtly can help build rapport.
- **Be mindful of your tone.** Speak in a friendly and welcoming tone of voice avoiding sounding harsh or monotonous.
- **Take the initiative to approach others first.** This shows that you are interested in having interactions with them.
- **Being genuine and authentic is key to being approachable.** Be yourself and show genuine interest in others. Show respect for others' opinions and perspectives even if you disagree.

- **Avoid being judgmental or dismissive.** Instead, be curious and appreciate that they think differently from you.

By incorporating these practices into your interactions, you can create an approachable aura that encourages others to engage with you.

OH, WHAT FUN!

Many workplaces organize social events, such as team lunches, happy hours and team building activities. These are the perfect relaxed setting to connect with coworkers outside a working office environment. If your workplace doesn't provide these opportunities, team projects are always a great way to foster relationship growth. The camaraderie and trust built up by working toward a common goal can create healthy, long-lasting connections.

LET'S TALK

It can be easy to join others in negative chatter in the workplace. Bad-mouthing a problematic manager, gossiping about a co-worker—it sure can help make the hours go by! However, these types of conversations close off chances for connection, rather than open them. If we instead remain positive, we will find that we have more opportunities to connect. People are drawn to happy, positive people; positivity is contagious, after all, and can even help improve our work environment.

Inviting a coworker out for coffee or lunch is a great way to begin a friendship. Or, if this is too high a hurdle at first, you can always drop them an email to check in occasionally and to foster friendships. Going for a walk together during a break is another great way to enjoy connecting.

PART II

HOW TO BE YOUR OWN BEST FRIEND

"When you make friends with the present moment, you feel at home no matter where you are."

—Eckert Tolle

BREAKING NEW GROUND

We now understand the impact that friend connections have on our happiness and quality of life, so we will concentrate here on deepening the connections that we have made. This involves preparation and planning. We can nurture our friend connections by investing in them in several different ways. One example is regular communication to keep in touch through calls, messages or meetups. This type of consistent communication helps maintain closeness. We can show genuine interest in our friends by asking about their lives and listening actively as they share details with us. This strengthens the emotional bonds of friendship. Being supportive is an important connection builder. Offering help, encouragement and empathy when our friends need it is a supportive way to build trust and deepen friendship. This prepares us to grow together with our friends and is a conscious way to improve connection.

This section will teach you many ways to consistently practice helpful behaviors to deepen your friendships and create meaningful connections that last.

MAKE A PLAN

The relative ease or difficulty with which we navigate life sheds light on our style of connection. For some, connecting is easy, even automatic—like breathing. For others, it's not so simple and takes a lot of work and focus to achieve…unless we have a plan.

However, regardless of the difficulty involved, connection is essential for maintaining balance in our lives. Even if it's a challenge, we must decide the best way forward with the task. All the same, many of us will read the research supporting the need for friendships and the impact that they have on our mental and physical health and *still* will not take the steps to bring healthy connections into our lives.

SMALL CHANGES, BIG CHANGES

It's surprising how quickly small changes add up to make a large difference in our life. This is to our advantage; as humans, we respond better to small changes than big ones and can commit to making them happen more easily and more consistently.

If we can commit to making one small change each week in our behaviors toward connection, over time we will make great strides. Here are some things to work on that can help build strong friendships in the future:

GIVE AND TAKE

Strong friendships are founded on reciprocity: both sides benefitting from the connection through mutual support. But reciprocity means more than just both sides being there for the other in times of conflict.

Reciprocity in a friendship also includes:

- Having shared interests and curiosity in one another's lives
- Taking turns setting up play dates and experiences to enjoy together
- An understanding that friendships grow, evolve and change shape over time
- Sharing feelings with one another, with integrity
- Being reliable and trustworthy in commitments

NOTICE YOUR WEAKNESSES

When I have new clients come to therapy and say they want to make a positive change in their lives, I start by holding up a mirror and help them to see exactly what will need to be different in order for those changes to take effect. Some accept the challenge; others run away.

It can feel overwhelming to let go of old habits, even if they are not good for us. We all have an innate tendency to be defensive, but when we become aware of how these habits hold us back, we can learn to let go of the defense mechanisms that cause disconnect in our lives and enjoy more connection. Unless we're brave and muster the courage to look inward, we can't determine what needs to change and we can't enjoy the benefits of self-improvement.

MAKE FRIENDS WITH FEAR

"Do one thing every day
that scares you."

—Eleanor Roosevelt

Every single one of us has moments where we avoid connection due to the possibility of rejection. Fear of rejection is the most common reason we don't reach out to others and unless we learn how to overcome that fear, we will miss out on valuable friendships that could last a lifetime.

Part of overcoming this fear is to make friends with it. Consider what this fear has come to teach you. How will overcoming this fear help you to connect with other people and live a healthier life? We *can* learn to use our fear as motivation to grow. Remember: no one ever grows during a period of comfort; growth comes from conflict.

BREAKING CYCLES

When we are in a difficult place in our life with seemingly no way out, we should first look to our connections. Are we isolating and growing our anxiety and depression? Or are we out in the world, engaging with our friends and enjoying life in connection with others? Odds are, we will find we need to reconnect. This need gives us the motivation to move forward, breaking the unhealthy cycles we might have picked up along the way. An example of an unhealthy cycle we might need to let go of and replace with a healthier pattern would be consistently avoiding meaningful conversations or emotional intimacy with our partner or close friends. This likely leads to emotional distance and a lack of understanding in relationships. For instance, if someone consistently withdraws or shuts down during discussions about important issues it can prevent resolution of conflicts and erode the connection between them over time. This pattern of an unhealthy cycle can weaken the bond and hinder the growth of a healthy and supportive relationship.

CONNECT WITH SELF

Do you find yourself constantly scanning the environment to understand the dynamics going on in a room? The only person you really need to monitor is *you.* What is going on for others is not about you; learn to let that go and just do you! Connecting with oneself involves fostering a deep understanding and acceptance of who you are.

Here are some ways to strengthen your connection with self:

- **Self-reflection:** Set aside time to regularly reflect on your thoughts, feelings and experiences. Journaling can be a helpful tool for this.
- **Mindfulness and meditation:** Practice being present in the moment without judgment. This can help you become more aware of your thoughts and feelings.
- **Self-care:** Prioritize activities and habits that nourish your physical, mental and emotional well-being. This could include exercise, healthy eating, adequate sleep and engaging in hobbies you enjoy.
- **Explore your values and beliefs:** Take time to identify what is important to you in life and reflect on whether your actions align with your values.
- **Emotional awareness:** Pay attention to your emotions and learn to acknowledge

and accept them without judgment. This helps in understanding your emotional responses and needs.

- **Set boundaries:** Establishing boundaries with others helps you define your identity and protect your emotional and physical space.
- **Seek support:** Connect with trusted friends, family members, or a therapist who can provide guidance and validation as you explore your inner self.
- **Learn and grow:** Be open to learning about yourself through new experiences, challenges, and feedback from others.

By nurturing these practices, you can develop stronger connections with yourself which forms the foundation for healthier relationships and a more fulfilling life.

AVOID ABANDONMENT

By connecting with yourself before you connect with others, you establish a firm base from which to interact with those around you. Making friends with others requires first being a friend to ourselves. Once we become more aware of our friendship qualities and what we have to offer others, we can approach forming relationships with more confidence. We can only learn these things through self-knowledge, which means that spending a little time with yourself can make the friendship journey easier.

Abandonment causes resentment, and self-abandonment is no different. When we ignore our self and our own needs in favor of others, we *disconnect* from ourselves. We can't form meaningful connections with others when we lack such a bond with ourselves. Having faith and trust in yourself, doing hard things and listening to the quiet voice only we can hear—this is the opposite of self-abandonment. Your very first connection should be with yourself.

Of course, this is easier said than done. Our relationship with our self is just as complicated and multifaceted as our connections with others, if not more so. Connection must be built and maintained with our physical self, as well as our mental, emotional and spiritual self. Rest, recovery and sleep—these are excellent ways to connect with our physical health. Doing something that we find personally meaningful is a way for us to connect with our spirituality. Being curious and open to learning is how we maintain our intellectual self-connection. Embracing painful emotions connects us with our mental health. Only after we have satisfied these conditions and reached a healthy connection with ourselves can we really start to seek out connection with others.

SELF-SACRIFICE

When we think of connection, we think of life coming together, making itself stronger. However, to have authentic connections, sometimes there are things that must first die. To be a good friend we sometimes must die to self. We must realize that the point is not what we gain from a friendship but also what we contribute. We must choose "thee over me" for a healthy connection. Our current culture does not advocate thinking about someone else's needs above our own but there is great gain when we do this in healthy ways. For example, we do something for another altruistically and that is of course unselfishly. And yes, that's hard—but it's also refreshing and in the long run will make you happier. Just because something is hard, doesn't mean it isn't worth doing.

PRIORITIZE

Staying connected with our friends does not happen by accident. We must make a conscious effort to be connected and engaged with the people in our lives. While this does take time and effort to achieve, the secret is making sure our connections are worthy of that investment. Making friendships a priority in our life is wise and benefits us both mentally and physically. For example, we can make a point to check in on our childhood friend once a week to support her while she is caring for her sick parent. Or we can make a point to go to mahjong once a week and interact in person with friends. This builds healthy structure into our days and helps strengthen in-person bonds.

SOFTER, NOT HARDER

Connection doesn't *have* to be difficult. If it is, it might be worth investigating whether we are trying too hard—pushing and pulling where we might be better off letting things sit. We can afford to slow down; there's no rush to connect, particularly if all the extra effort is going to waste. Approaching people softly means listening to our heart and intuition and being fully present in our interactions—letting things unfold naturally like a blossom, rather than trying to force them along a preconceived path. Blossoms can't be forced into blooms; they take time to reveal their natural beauty. Friendships are the same.

Practicing friendship as a slow unfolding process involves nurturing connections over time in a deliberate and meaningful way. We can do this by being present and genuine. Showing genuine interest in others by being fully present in conversations and listening actively and thoughtfully.

Building trust is a gradual process and it takes time to develop. We can share bits of personal information gradually as trust builds and respect others' boundaries as they do the same. Again, not rushing the process. By approaching friendships as a slow unfolding process, we can cultivate meaningful connections that are built on trust, understanding and shared experiences.

SHARING

"To keep your secret is wisdom but to expect others to keep it is folly."

—Samuel Johnson

Sharing information about ourselves is one of the keys to forming and strengthening connections. That said, the more intimate the information, the more careful you need to be in choosing who to share it with. Before sharing, stop and consider: "Is this someone I want to share my most sacred experiences and thoughts with?" Contemplate what they might do with the information you are going to share with them. Do they seem like a safe person to trust with your secrets?

Most importantly, ask yourself why? "Why do I feel a want or need to share with this person?"

For example, everyone doesn't get to know all of our business. It's important to discern if the person we are contemplating sharing our story with is safe? Are they capable of holding space for us and being understanding and empathetic?

GRATITUDE

"This is a wonderful day. I have never seen this one before."

—Maya Angelou

Be grateful for each day and the new opportunities it offers for connection with other humans. If we begin our day in connection with our higher power with gratitude for the new day, then we will find little pieces of sunshine scattered throughout our day. Research supports the idea that gratitude balances our mental health when our brain releases dopamine and this is a chemical release that makes us feel happier. We feel more like connecting and interacting with others when we are happy. The sunshine we might find is healthy interactions with other humans scattered throughout our day. It can be a lovely surprise to run into an old friend at the bookstore and to reconnect.

YES AND NO

Let's reflect on all that we have learned up to this point. This review can aid us by reminding us what we have learned from the content thus far. Are you able to identify areas of your life that you might need to change in order to build healthy connections with others? What do you need to improve upon based on this reflection, and what are some things that you may already have mastered? Can you say, "YES!" to connection and "NO!" to unhealthy patterns that bring about disconnects?

Some NO patterns look like:

- Too much time online
- Self-isolating
- Having no sense of adventure
- Rigidity in opinions and habits

Examples of YES patterns include:

- Practicing active listening
- Answering messages from others promptly
- Volunteering to help out in one's community
- Openness to new ideas or experiences
- Extending invitations to others

PART III

HOW TO MAKE NEW FRIENDS

"The only way to have a friend is to be one."

—Ralph Waldo Emerson

WHAT MAKES A FRIEND GOOD?

What are the characteristics of a good friend? Being a good friend means being supportive, trustworthy and understanding. It means being an active listener, being present in both good times and bad, offering help and encouragement when it is needed. Respect, honesty, empathy towards your friends' experiences and feelings—these are all elements of a good friend. Good friends maintain open communication and show appreciation for their friend's presence in their lives. Take a moment to recall the best friend you have had. What did they do to be such a good friend? How can you incorporate elements of their behavior into your new and current friendships?

FRIENDSHIPS REQUIRE CULTIVATION

"Remember, the most valuable antiques are dear old friends."

—Jackson Brown

How much time do you invest in your friendships? We spend lots of time on our jobs, our homes, our bodies, but when it comes to our relationships, we have a tendency to shortchange our time investment—especially during times of turmoil or stress. We know our friends are an important element of a happy life. Shouldn't we spend more time creating friendships or nurturing the ones we have? Remember, time spent with a friend is never wasted.

BE TRUSTWORTHY

"All the world is made of faith, trust and pixie dust."

—Peter Pan

Trust is *the* cornerstone of real friendship. We all want a friend we can trust, and we all want to be a trustworthy friend. By being reliable and honest—always doing what we say we will do—we form the foundation for a meaningful friendship to grow and flourish. Building and nurturing trust and friendship requires communication, empathy, and a willingness to invest time and effort. All of this together is a recipe for trust and creates the loving connection called friendship.

BE LOYAL

"True friends came into your life, saw the most negative part of you, but are still not ready to leave you, no matter how contagious you are to them."

—Michael Johnson

What does it mean to be loyal to one's friends? It entails showing up for one another through thick and thin, supporting one another's dreams and endeavors, and staying true even when faced with challenges and disagreements. It's about trust, honesty, and commitment to your friend's well-being and happiness. For some people, loyalty is *the* most important element of friendship. What is the most important element of friendship for you? What do you need in order to feel securely connected with someone?

BE REAL

"Be yourself. Everyone else is already taken."

—OSCAR WILDE

As we've discussed, connecting with our authentic selves is vital to being able to connect with others. A healthy friendship cannot be built on a shaky foundation, and without understanding who we are, what we're looking for in a friendship, and what we have to offer in return, any connection we create will be weak or incomplete. Being authentic involves embracing who we know ourselves to be, being aware of our values and expressing our genuine thoughts and feelings without pretense. It's about connecting with vulnerability and honesty in our interactions with others.

Make a list of the elements you call "you." Include the things you value most. This will show you what you need to begin an authentic connection.

BE VULNERABLE

"Meaningful connection requires meaningful work."

—Brene Brown

There is a scary concept: *be vulnerable.* Even the mention of it is enough to make some people run for cover. We have all had an experience where we were open and vulnerable with another human and in the end, we were betrayed or mistreated.

I tell you now, *connect anyway.* You may end up having another negative friend experience; you also may not. But even our bad experiences with friends are important because they teach us something…even if that something is simply, "How not to behave in a friendship." It is possible to learn things like being genuine, being honorable and showing up for our friends by observing their lack in others.

Even when it's hard to be vulnerable or we don't feel like we're up to it, do it anyway. True friendship and authentic connection cannot occur when one side is closed off; there is no real friendship that isn't a two-way street.

BE THERE

"Lots of people want to ride with you in the limo but what you want is someone who will take the bus with you when the limo breaks down."

—Oprah Winfrey

I once had a client come to me because, after she had made the decision to leave an abusive marriage, almost everyone in her life had walked out on her. They chose to let their own beliefs block them from showing up for this person when she needed them the most. They chose to disconnect from her at a time when she most needed connection and closed the door rather than sit with her through the messy reconstruction of her life. In essence, she came to me looking for exactly what her friends should have provided: support as she built a new life. It's sad to see how unkind and self-serving other humans can be when their judgment drives their decisions.

What better time to be there for another human than when they need you most? True, this may often require letting go of prejudices or judgments about how you think something should be handled. You may not agree with all of the choices that brought your friend to their current predicament. But all of this is irrelevant; what matters is being there, showing up in the sunshine *and* the rain. My client survived her ordeal and found new, real friends—all while learning a lot about friendship and human nature.

Be present in your friendships always, but especially during the storms. True friendship shows up in the hard times and through them, the strongest connections are built. If you *can't* do that, then explore that side of yourself. Look at yourself in the mirror and ask, "Why? What is holding me back, what about me needs to change before I can be a true friend to those who need me?"

BE DEPENDABLE

"The greatest ability is dependability."

—Bob Jones

Our friends need to know that when we say we are going to do something, we will do it. Our word needs to be golden. Dependability is a crucial aspect for secure connections. A friend who consistently delivers the expected results, in all sorts of conditions, is dependable. When I know I can count on you as a friend, that is irreplaceable.

Being a trusted friend is important and involves honesty, confidentiality and integrity but being a friend other friends can depend on is also essential. A dependable friend is consistent and reliable in their actions and can be counted on to follow through on commitments and promises.

A dependable friend is supportive. They offer assistance and help when needed, showing up for you in practical ways. Consistency in a dependable friend looks like steady and predictable behavior, making them someone you can rely on in various situations.

In essence a dependable friend is someone you can rely on for support and consistency while a trustworthy friend adds honesty, integrity and ethical behavior to the relationship. Both qualities are crucial for strong lasting friendships, but these emphasize different aspects of interpersonal reliability and integrity.

BE UNDERSTANDING

"One of the most beautiful qualities of true friendship is to understand and to be understood."

—Seneca

As discussed, our friends aren't always coming to us on their best day. They might be stressed from work or family situations; they might be angry, frustrated or anxious. Being a safe place for them to land is essential to the growth of deep friendships. Caring is kind and serves as fertilizer in the garden of friendship.

BE HAPPY

"Friendship is a treasure trove of joy, where the currency is not material wealth but the priceless exchange of laughter, understanding and unconditional support."

—Anonymous

There are many studies correlating happiness with friendship connections. Arthur Brooks, a Harvard expert on happiness, says friendship connections are one of the four pillars of happiness. Using the widespread loneliness that came about during the coronavirus as evidence, he directly correlates happiness and friend connections with well-being. He also believes that creating connections with other humans in our lives is how we build a happy life.

Friendship brings about myriad positive emotions. It creates joy, laughter, understanding, trust, companionship, a sense of belonging and a feeling of being supported. These are the physiological and emotional benefits which improve our wellbeing when we engage in healthy friend connections according to the World Economic Forum and many other research findings. Our bonds with others enrich our lives with warmth and fulfillment, leaving us feeling appreciated, valued and loved.

BE ALL EARS

"When people talk,
listen completely. Most
people never listen."

—Ernest Hemingway

When you are with your friends, do you listen in order to form a response, or are you really paying attention to what they are saying? Are you only hearing the words they're saying, or are you soaking in their message as well? Are you interested in the details of what they are sharing or simply nodding and thinking about your pickleball game?

We don't have to share our thoughts with everyone. Determining how much to share and who to share it with is a part of managing our various levels of friendship. Other times, the people we are connected to simply want to talk and aren't looking for any advice from us; they are just processing their thinking out loud. So, before offering advice, it might not hurt to get into the habit of asking, "Are you asking for my advice or do you just need me to listen?"

Some tips for listening include:

- Practice patience and realize that sometimes silence is powerful.
- Allow the person talking to think and process without rushing them to fill gaps.

- Avoid making assumptions by focusing on what is being said without jumping to conclusions or making judgments prematurely.
- Ask clarifying questions if something isn't clear
- Ask open-ended questions to gain more insight like, "Can you elaborate on that?"

Being all ears is essential for nurturing authentic relationships, but this type of active listening requires proper communication skills. Believe me: it's easy to tell when someone is sincerely paying attention to you, when you are *really* being heard. Invest in your friendships by being all ears when your friends are sharing important messages with you.

Admittedly, it is becoming more difficult to do this, to maintain a laser focus on the other person due to having so many tabs open in our minds. Our focus has shrunk and become compromised due to excessive digital device usage. But I believe active listening represents an exercise in caring, in making an investment into our relationships leading to more solid connections.

BE EMPATHETIC

"Empathy is the bridge that connects hearts, allowing us to walk in another's shoes, feel their pain and celebrate their joy as if it were our own."

—Unknown

Empathy is the ability to understand and share the feelings of another person, allowing us to connect with them on a deeper level. Empathy in friendship is truly the invisible thread that weaves hearts together, allowing us to understand each other's joys and sorrows without judgment—to walk hand-in-hand through the highs and lows of life. There is no greater joy than sharing with a friend and having that, "Me, too!" connection. When you get me, when you've had the same experience that I am sharing with you, it creates a special connection between us. This is the very best part of friendship.

There are several ways we can practice empathy in our friendships, and they involve actively understanding and sharing others' feelings. We can do this through active listening and paying full attention to others without interrupting. Very hard to do sometimes but if we focus on understanding their words and emotions, we will be successful. We can also put ourselves in their shoes. We might imagine how the other person feels based on their experiences, perspective and background. We might do this by asking open-ended questions to encourage others to share their thoughts and feelings. To them, this likely feels like we are showing genuine interest.

We can validate feelings by acknowledging and respecting others' emotions, even if we don't agree with their perspective. Practice nonjudgment by avoiding jumping to conclusions or making assumptions. We must be open to different viewpoints. Reflective responses occur when we repeat back what we hear to confirm understanding and show we are engaged. By consistently practicing this strategy, we can develop empathy as a skill, improving our relationships and understanding of the world around us.

BE POSITIVE

"For without good friends, no one would choose to live, though he had all other goods."

—Aristotle

Remember, we all have warts. When making friends, try not to concentrate on their imperfections. All humans are imperfect—it's part of what makes us human. We each have to learn to overlook small negatives and focus on the positive aspects of the other person to enjoy long lasting friendships. Part of having a healthy and happy life is having a short memory!

Our friends are imperfect—so love them anyway. Even though their flaws may look different from ours, seeing the ugly side of another human is no excuse to disconnect, unless their faults are harmful to us. Otherwise, learn to enjoy the differences between us. We stand to learn the most from those who are the least like us.

BE RESPECTFUL

"Respect is the acknowledgement of each other's worth, the appreciation of differences and the commitment to treat one another with dignity and kindness."

—Unknown

Mutual respect is the building block upon which trust, understanding and admiration are built in a friendship. Respect is a non-negotiable in secure relationships. Mutual respect means recognizing each other's worth, boundaries, and differences while fostering a relationship of equality, appreciation and lasting camaraderie. To have a relationship without this would be like a day at the beach without the sun!

BE WILLING TO SAY "I'M SORRY"

"Apologies are the language of reconciliation, spoken with sincerity and humility, paving the way for forgiveness and restoration of trust in the bonds of friendship."

—UNKNOWN

In friendships, apologies are the glue that repairs the cracks that inevitably form in any relationship. We are all different people and conflicts with one another are inevitable. The mark of a true friendship is one that can survive this type of everyday wear and tear, and apologies are critical in doing so. By restoring trust and strengthening the bond between us with humility, sincerity, and a commitment to do better, apologies are not just words: they are a profound acknowledgement of our mistakes, the humble acceptance of responsibility for our actions and the courageous commitment to repair what was broken, forging a path towards healing and deeper connection.

BE ALTRUISTIC

"The most I can do for my friend is to simply be his friend."

—Henry David Thoreau

Do not lend another your help with the expectation that they will one day return the favor. Give without strings attached. If we don't give altruistically, then we harm the relationship and undermine trust. Genuine acts of kindness become transactional, and friends feel manipulated and used when they realize there were hidden expectations. It also creates imbalance as it disrupts the natural balance of reciprocity in friendships. Genuine friendships thrive on mutual support and care without keeping score or expecting immediate returns. The strain on the relationship leads to resentment or conflict if the friend feels pressured to reciprocate in ways they may not be comfortable with or able to fulfill. In essence, friendships should be nurtured through genuine care and support, rather than expecting something in return for every act of kindness. True friendships flourish when there is mutual respect, trust and the freedom to give without expecting immediate reciprocation.

PART IV

HOW TO NURTURE FRIENDSHIPS THAT LAST

"A real friend is one who walks in when the rest of the world walks out."

—Walter Winchell

KEEPING CONNECTIONS HEALTHY

We can all likely boast that we have some connections in our lives with other humans. Hopefully this book is motivating thoughts around the depth of these friend connections and the meaningful ways discussed here in strengthening them have been helpful for you, the reader. Deepening connections with others often involves intentional effort and genuine engagement. Let's face it, it can be hard work, but it is also worthwhile. In our modern world we can begin to feel lost or diminished in terms of human connections. It's easy to feel isolated or disconnected from others despite being more digitally connected than ever. Finding and nurturing human connections is challenging, but we know it remains crucial for emotional wellbeing and personal fulfillment.

There are myriad ways that we can deepen and strengthen our friendship connections.

We can do this by asking thoughtful questions that go beyond small talk to explore deeper topics that are important to us both. We can practice respect in terms of our friends' boundaries and honor their personal space. This shows consideration and respect for their individuality. We can also show appreciation for our friends' qualities and show gratitude for their efforts in the friendship.

This section will provide meaningful ways to deepen, build and sustain more meaningful and fulfilling connections with others in our lives.

HOW CAN WE MAKE OUR CONNECTIONS DEEPER?

"There are two ways to be happy: improve your reality or lower your expectations."

—Jodi Picoult

Perhaps we already have important connections, but we want them to be a bit more secure. We want to feel safe with the other person so we can more easily be vulnerable; or perhaps we want a connection where we can embrace humor, playfulness and fun together.

There are three things we can do to deepen our relationships:

1. Be curious with the other person, eager to learn more about their thoughts, feelings and needs.
2. Adjust our expectations for the friendship as needed so that they are realistic.
3. Express our needs, thoughts and feelings so that our expectations are clear.

We have learned in this book so far what good friend connections look like. We know they involve intentional actions on both parts in the friend duo in order to maintain safety, trust and understanding. We have learned the elements of friendship that bring us closer and where to find these friends. Some are already in our friend group, we simply have to nurture those relationships to grow deeper and we've found we can do this through sharing our thoughts feelings and experiences openly. This is what it looks like to be vulnerable. But as we grow comfortable doing this, we find that the comfort of communicating with dependable friends regularly outweighs the fear of being vulnerable. This provides a sense of belonging and emotional safety that is invaluable.

SLOW DOWN

"Let me give up the need to know why things happen as they do. I will never know and constant wondering is constant suffering."

—Caroline Myss

Relearning the art of connection and finding ways to slow things down are currently high priorities in people's minds. Our world is fast-paced and filled with constant distractions—far from the environment humans were designed for. We are not our best selves when we are rushing through our days, our weeks and, ultimately, our lives. It's easy to lose sight of real connections when we live at breakneck speeds. Living in the present moment seems all but impossible.

Slowing down is the art of living in the present moment. To live today fully, we must let go of what has come before. Disconnect from yesterday—it's over and it's not coming back. But today is here, *now,* and we can still make it the best day of our lives so far. Connecting with the present and disconnecting from the past is another lost art, but one well worth re-learning. You will never be able to move backwards and relive those days; it's not possible, but that won't stop you from missing today while trying to relive yesterday. Rather than wasting the time you have been given, choose to make the forward journey the very best you can.

Rediscovering a slow, more deliberate and more realistic pace leads to greater fulfillment, deeper relationships, and a more meaningful existence overall. What are some ways you can try to incorporate slowing down and making real connections in your life? I'll share with you a few suggestions that I find helpful.

Consider trying the approach of prioritizing quality time with friends over simply fitting them into an already overloaded schedule. Use this time to engage in activities that foster deep conversations and shared experiences. Make time with friends a time to be all in and present, offering your full attention. Be mindful of your pace during friend time. Recognize when you are moving too fast and need to slow down. When we balance our personal time with our social life, it ensures we don't become overwhelmed and overloaded. A slower pace helps eliminate anxiety during friend time and you will both leave your time together feeling refreshed and restored. Think of friend time in terms of "I get" to enjoy time with my friend rather than "I have to."

The Round Table

So many wonderful connections happen at the family table. One family meal together each day results in more stability and emotional maturity in children, as it teaches kids the important skills of socialization. Surely we can all think back to lovely times spent with our families gathered around the table, connecting and engaging.

CHECK ON YOUR FRIENDS

Connection relies on contact and communication, especially when it's unplanned and unannounced. Make the call. Write the note.

Send the text or drop by for an unexpected visit. Remind your friends that they are important to you and that they have been on your mind.

Make a point to have consistent contact with your friends. Actions over time become habits and habits become who you are. Success doesn't come from what you do occasionally, it comes from consistency—and reaching out to your friends is a great way to form successful connections.

RETURN MESSAGES

"If you actually want a meaningful life, pursue love. Do selfless acts of love for people."

—Bob Goff

There is nothing worse than feeling ignored. Make it a point to return messages, be it an email, a text or a call, before the end of each day. Don't let the sun to go down on a message that is not returned. Over time, this habit will improve the connection you feel with your friends. It is a small gesture with long-lasting effect and is the difference between being a friend and being an incredible friend.

WHAT'S IN A NAME?

Guess what? We all enjoy hearing our name. Small efforts can yield big benefits, and the tiny habit of using a person's name in conversations with them can endear them to you for a lifetime, all because they'll know they were important enough to you for you to remember their name. So, give it a try—it's a small gesture that means a lot.

MOVE YOUR BODY

"I find each day too short for all
the walks I want to take and
all the friends I want to see."

—John Burroughs

When is the last time you invited a friend for a walk? This small, shared moment kills two birds with one stone: you are fostering friendship and connection while moving your body and working towards optimum physical health. Walking may seem boring alone, but the time will fly by if you have a friend to chat with while getting your steps in. For example, I walk six days a week and each day I invite a different friend. For me, it was as simple as replacing my lunch gatherings with walking. It is healthier in every way—and leads to a lot more socializing!

LAUGH TOGETHER

"Laughter is a very good index of the people we are with."

—Sophie Scott

Laughter truly is the best medicine. We all love to laugh; we all *need* to laugh. Laughter draws us to one another while also improving our emotional and physical health. Did you know laughter can strengthen our immune system, uplift our mood, diminish pain *and* decrease our stress levels? No wonder we all gravitate towards people who make us laugh!

The Golden Girls

Who still remembers this show? *The Golden Girls* was a sit-com in the 80s about a group of women enjoying their "golden years" together. There was always an obstacle to overcome, but through those experiences, the connections the characters formed were unbreakable, irreplaceable. They shared with and looked out for one another; they built community together; they grumbled and giggled and these emotions drew them closer together. Their everyday life was talking and teasing and taking on the world together. What a brilliant idea!

PLAY TOGETHER

Shared recreation is the best way to build a friendship. Through play, we develop memories that bond us with our intimate friends. The bonding chemical oxytocin, which gets released as we play, brings us internal joy. It's an important part of the science of connection and how it affects the body and brain.

People who don't have friends to play with report a poorer quality of life and an increased risk of premature death. Thankfully, friendships can be made and fostered at any age. In other words, we are never too old to meet new friends and play!

Not Just Happier Together—*Better* Together

I was visiting with a girlfriend the other day. We were walking the length of her swimming pool and talking about the reasons we enjoy meeting in person. We have a close friendship, and it is because of the little things that happen when we meet in person. She picked up a towel to dry herself off after our pool time and also picked up one for me. I brought her an after-exercise snack when packing one for myself. These little niceties are what bond us, one to the other. I smile when I think about the ease I feel when being together and catching up. My heart is warmed by the thoughtful gestures of in-person meetings.

SHOW YOU CARE

"Each friend represents a world in us. A world not born until they arrive, and it's only by this meeting that a new world is born."

—Anaïs Nin

The important friends in your life will become obvious with time. As you share experiences with them, you will instinctively know who you can trust and be vulnerable with, because they will stand out head and shoulders above the others. When sharing your innermost thoughts and feelings, a good friend will guard them like gold. They will care...and you will, too.

A few tips for showing you care could include:

- **Small acts of kindness** such as remembering important dates, surprising them with something they like or offering words of encouragement can make a big difference. This is what being thoughtful looks like.
- **Using positive body language** by maintaining eye contact, nodding in agreement, and using open inviting gestures. Your body language should convey warmth and attentiveness.
- **Remembering important details** by keeping track of key events, preferences, or personal details they've shared. Mentioning these in conversations shows you value them.

"Man's friendships are one of the best measures of his worth."

—Charles Darwin

ONE LAST THING

You've reached the end of this little book on big connections. I truly hope these tips and insights for connecting with others help you to find new friends and make solid connections. Most importantly, I hope you realize connecting is the essence of a life well lived. Everyone needs and deserves to have healthy connections to add beauty and meaning to their lives.

Yet according to a study, one in four humans report having no one in their life they feel connected with. In truth, we can all benefit from strengthening our connection muscles. We must realize we are not alone. Connection is our friend, one which makes us healthier and happier individuals. When we can love and connect with others through our joys and our struggles, we will have less depression, less anxiety and less disconnect. Our world will be a better place.

We are not alone. We are *all* interconnected. We may have lost the art of connecting with one another, but it's an art we can relearn—reestablishing strong connections that will improve our lives for years to come.

Reviving the ability to make deep connections involves relearning how to have face to face interactions to cultivate deeper relationships. By incorporating the practices shared in this book we can enhance our ability to connect deeply with others and breathe life into personal interactions that build strong relationships. To do this we have to leave the sidelines of life and jump in with both feet. Jump into action and into connection. Put the phone down in the line at the coffee shop and engage with those around you. Leave the bench at the park where you sit endlessly scrolling and watching kids build a snow man, join in and feel the connection as you help create the snowman.

Show up and create human connections. Embrace the opportunities that present themselves to you every day to be connected and engaged and to enjoy the happiness and fullness of life.

"I have learned that to be with those I like is enough."

—Walt Whitman

RESOURCES FOR CONNECTION

Cope, Stephen. *Deep Human Connection* (California: Hayhouse LLC., 2019)

Whatley, Lori. *Connected and Engaged* (New York: Hatherleigh Press, 2024)

Yanagihara, Hanya. *A Little Life* (New York: Doubleday, 2015)

"Friends are the siblings
God never gave us."

—MENCIUS

A NOTE FOR PARENTS

As parents we can help our kids understand the importance of being a good friend. We can have conversations about good friendships with them and model good friend behavior for them to see and learn from. It will assist them in their friendships as they navigate life and learn the importance of connecting with others to create lifetime bonds. We can share with them all the healthy connection habits discussed throughout this book.

For example, if we see our kids are often texting rather than having in-person conversations, we can encourage them to use texting to begin a conversation about getting together in person. Text the logistics for the meet up and then offer to help them choose a place and time to meet and possibly work with the other person's parent to make this successful. When you put together your child's snacks for school offer to add a snack for them to share with a friend. This is a productive way to foster friendship and practice healthy in-person connection.

Other good practices to get into include:

- Sign kids up for in-person group activities like team sports, where they will interact face to face with peers.
- Set boundaries on kids' screen time and use will encourage them to instead have more in-person activities.
- Allow your kids to see you having in-person conversations and activities to form connections with friends.
- Participate with your kids in community events where they will meet in person with friends and their families.

By creating opportunities for face-to-face interactions and guiding them in social situations, you can help your kids build lasting friendships beyond texting.

ACKNOWLEDGMENTS

Always, I want to thank the ones who you don't see but for who, without their encouragement and guidance, we would not have this book.

Thanks to my friends who helped me make this message clear. They read and reread chapters and sections and gave me honest and helpful input.

Thank you to the professionals who developed this idea. Now, a book is born. Ryan, Ryan, Hannah and Andrew: you each have an incredible eye for what connects us.

My family is my inspiration. To make the world a peaceful place for them is my goal.

"When the world is so complicated
the simple gift of friendship
is within all of our hands."

—Maria Shriver

ABOUT THE AUTHOR

Dr. Lori Whatley is a licensed Marriage and Family therapist with a Doctorate in Clinical Psychology. Based in Atlanta, Georgia, she specializes in healthy human connections and believes connection is the cure for the loneliness epidemic in our world. She helps people connect in a world full of disconnection. Lori travels and enjoys speaking to groups about the connections that enhance their lives. Over the three decades of her career, she has written three books about connection, including *Connected and Engaged: How to Manage Digital Distractions and Reconnect with the World around You.* Through her work, Dr. Whatley has helped thousands of clients improve their connections and, in doing so, improve their lives.